Deep Time

Melaina Faranda

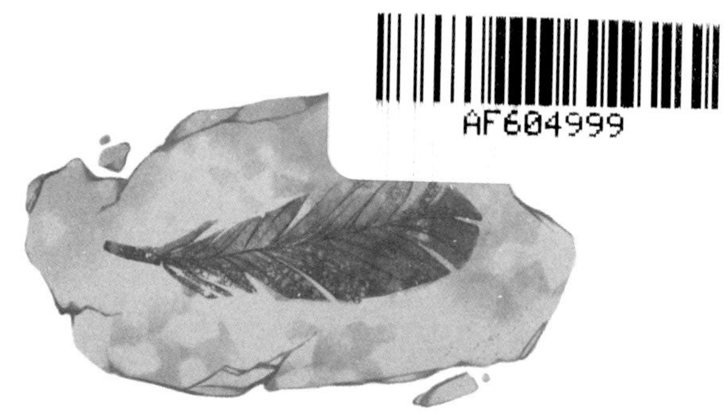

Contents

One

The Diabolical Duo

Like clockwork, one of the twins started crying. Then the other began. Jack pulled the pillow over his head. It was no good; even a closed door, plush wool carpet and being upstairs were no match for their combined lung power.

Groaning, Jack threw back the covers, causing a mild meow of protest from Skittles, who was now half buried under a quilt. Jack stumbled to the wardrobe for his running clothes. Then, silently opening his bedroom door, he listened to check the coast was clear before creeping downstairs. At the bottom step, he trod on a toy truck with a loud crack. Jack froze before tiptoeing to the front door.

"Jack, is that you?" called a tired voice from another room.

Jack was tempted not to answer. The front door was within arm's reach. Jack could almost smell the dewy grass and see the morning sunlight combing fingers of mist through the neatly trimmed trees that lined the street. Freedom was only steps away.

"Jack!" This time, Mum's voice was sharp with frustration. "Could you please come and give me a hand with Daisy while I deal with Danny?"

Jack trudged towards the twins' room. Stuffed animals littered the floor. Daisy jumped on her bed as if it was a trampoline while Danny wailed and flailed his fists as Mum tried to change him.

Jack caught Daisy mid-bounce. "Come on, Daisy, how about we go find Skittles?"

Daisy beamed. One of her favourite activities was terrorising the poor cat with her clumsy, full-body toddler hugs. "Diddles hiding," she said.

"Yep," Jack replied, inhaling her apple-soap, little-kid smell. "Skittles is trying to survive."

Mum frowned as she gripped Danny tighter. "Yesterday, that cat tried to bite Daisy. If he's going to be aggressive, we'll have to find him another home."

Jack felt a sting of injustice. Skittles was *his* pet. He had been given the ginger kitten shortly after the twins were born. Maybe it was some sort of compensation for Mum and Ryan paying almost zero attention to Jack since bringing home the diabolical duo, as Ryan sometimes called them.

"It could be because Daisy's version of hugging is basically strangling," Jack muttered. But Mum wasn't listening. She never seemed to listen. Heat flooded Jack's face at being used, once again, as an unpaid babysitter.

He plonked Daisy back onto the bed, and she instantly scrunched up her cheeks. "I wanna play with Diddles!" she wailed.

Mum put her hands over her eyes and groaned. Her dark hair had already been tugged loose from the smooth bun she wore to work, and she looked even paler than usual. Her dressing gown slipped to show her work clothes beneath. She couldn't risk the many ways the twins could smear her clothes with their perpetually grubby paws, so she'd put her dressing gown back on after getting dressed. "Jack, pick her back up," she said.

"No."

"Do as you're told, please." It wasn't a request.

Jack ignored the warning tone. “No. I’m not your babysitter. And why isn’t Ryan helping here, anyway?”

“He’s looking for jobs on his phone, Jack.”

Jack rolled his eyes. “That’s what he says. Why can’t he do it after you’ve gone to work?”

Mum sighed. “He’s struggling at the moment, that’s why. The company collapse has been a terrible time for him; it’s a big blow.”

Once again, his mother seemed to consider everyone but Jack. “And you’re not struggling?” Jack demanded. “Or me? Why can’t I even just go for a run without having a kid dumped on me? You know what? I would way prefer to keep Skittles and give the twins away. At least Skittles doesn’t trash the house and scream all day long. Everything got worse when they came along.”

Mum’s dark eyes flashed darker. “Take that back, Jack. You know you don’t mean it.”

“Yes, I do. They have made being at home horrible.” Jack headed for the front door.

“Don’t you dare walk out, young man!”

Jack rounded on her, too angry to control what came out. “I’m sick of you treating me like a servant!”

His mother recoiled and, for once, even the twins fell silent, both gazing at him with wide eyes.

"I didn't choose for you to have the twins!" Jack shouted. "I didn't get to choose *anything*. I didn't ask for you to leave my dad. I didn't even get a say when you and Ryan got married."

Jack shuffled his feet, unable to meet Mum's eyes. They both knew that Ryan had always been good to him.

"If you think it's so terrible here, maybe you should go and live with your father," Mum snapped.

"I wish I *could* just go and live with my real dad!"

Mum's eyes narrowed. "If that's what you want, then fine. See how much better you like it. I'll call him today."

"Good." Ignoring the sudden howls issuing in stereo from the twins, Jack slammed the front door and stormed out into the street.

Two

I Think You're Right

Jack knew he was going to be in trouble when Mum got back from work. He had spent his day at school fuming over their fight. She'd probably had a tough day, too, but he didn't want to think about that. Instead, Jack brooded over the unfairness of her suggestion of getting rid of *his* cat. And the way she made excuses for Ryan all the time, *and* how she expected Jack to do everything for the twins.

Not once in Jack's life had Mum ever said that he should go to live, or even stay, with his father, even when she was angry with him.

Trudging through the front door, Jack could hear the twins clamouring in the kitchen. Ryan must finally be out of bed and feeding them afternoon tea.

The kitchen looked like a toddler bomb had hit it. The twins had opened all the bottom drawers and pulled out what looked like every single pot and pan. A sticky yellowish trail snaked along the floor, and floury fingerprints dusted the chairs and table legs. Ever since Mum and Ryan had stopped being able to afford childcare, there had been constant chaos in the same house that used to feel welcoming.

Ryan flashed Jack a quick grin of relief as he peeled Danny off his leg. "Reinforcement has arrived! Hi buddy, could you grab a cloth for me, please? I'm hoping that yellow stuff is honey and not something worse."

"Buddy" was what Ryan had called Jack from the time he had moved in, when Jack wasn't much older than the twins were now. A lot of the stuff they had done together – going to play catch in the park, riding the scariest rollercoasters (while Mum waved, green-faced, from below), stand-up paddleboarding on holidays – now felt like distant memories. Jack's only value now was his ability to hold a toddler or pass something to clean up the twins' mess, or *worse* – clean it up himself.

Jack grabbed a dishcloth and resentfully scrubbed at the sticky stuff.

As Mum's car pulled into the driveway, the twins bowled towards the door, shouting, "Mama, mama!"

From the way she picked each twin up with a kiss and placed one on either hip, it looked as if she might be calm. Until she met Jack's eyes. She nodded curtly. "We'll talk after dinner."

***

When Mum and Ryan had both worked full-time, they sometimes ordered food boxes that came with all the ingredients in the correct portions for meals. It had been Jack's job to assemble the recipes on Tuesday and Thursday nights, the days when he didn't have football practice or his saxophone lessons.

Ryan was now the family cook, which meant they ate a lot of pasta. Jack missed the variety. He also missed helping Mum make Chinese treats: dragon's beard candy and the *tanghulu* apples her *lao lao*, or grandmother, had taught Mum to make when she was a little girl back in Beijing.

While Mum tucked the twins into bed, Ryan flopped on the couch and stared, glassy-eyed, at his phone.

Jack stacked the dishwasher and opened a tin of sardines for Skittles. He was still wiping the granite counter when Mum came in.

"We need to talk," she said.

"Okay. What is it?"

Mum ducked past him to get herself a glass of water. "I thought about what you said this morning, and I think you're right. It's not easy here right now, and you should have the opportunity to spend some time with your father. So, I called him."

Jack tried to conjure a picture of his dad. His recollection was of a tall guy with bright blue eyes taking him to the museum, where they spent a lot of time looking at boring rocks.

Jack's dad was a palaeontologist. Jack had once boasted to other kids in his class that his dad dug up dinosaurs. But that was years ago. Apart from an awkward annual phone call for his birthday, during which Jack would be itching to get off the phone, Jack had no other contact with his father.

Mum pulled the clip from her bun, letting her hair down. "He's agreed to take you. Apparently, there's a school near to where he's on a dig. You'll be flying out next weekend."

Jack's heart thudded to a stop. She hadn't discussed this with him. Saying he wanted to live with his dad was just something he said because he was angry with Mum.

He suspected this suited Mum. It seemed she just wanted to get rid of him.

"But –" Jack began.

He never got to finish. From the twins' bedroom came a piercing wail.

"Ryan?" Mum shouted. "Could you please go and check on the twins?"

Through a lull in their crying came an unmistakeable loud snore. Ryan was sprawled on the couch, his mouth open, while Skittles contentedly settled on his belly as if it was a comfortable cushion.

Mum balled up her fists, her knuckles whitening. She dug them into the sharp bones of her hips and rushed to the twins' room.

Jack considered how to appease her without losing his dignity. He wasn't going to throw a fit like one of the twins, but he also didn't want to live with a stranger who just happened to be his biological father. He was going to have to apologise to Mum, even though he didn't feel like he had done anything wrong.

When the screams finally quietened down, Jack heard Mum's voice floating through the window, from out among the remaining shrubs the twins hadn't yet flattened with their daily tricycle raids. She was in the backyard on her phone. From her hushed tone, Jack guessed she was talking to her best friend.

"I know, Sal, but he's nearly a teenager. He used to be so easy. It's just that with the twins being such a handful and Ryan falling apart, I can't cope with the attitude. Besides, it's about time he got to spend some time with his dad. It will give him some freedom and more independence."

Her voice lowered to a whisper. "Honestly, Sal, I give it a week before he asks to come back."

Jack froze. Now there was no way he could tell her that he wanted to stay.

Three

Rocky and Dusty

Jack turned up the volume of his music to drown out the plane's engine as he gazed at the reds, browns and oranges below – an endless expanse of baked desert, with the occasional stark shadow made by a rising rock formation or cliff.

The first flight had been on a typical passenger jet, but then Jack had boarded a tiny six-seater plane with a propeller on the nose that looked like something from one of those old-fashioned war movies.

His whole family had come to see him off, the twins straining in all directions before spotting the travelators and breaking free. While Mum chased them, as they scattered like marbles in opposite directions along the polished tiles, Ryan pulled Jack into a bear hug. "It won't be the same without you, buddy." He stepped away to peel open his wallet, and checked Mum was still busy herding the twins before slipping Jack a one hundred dollar note. "This is for emergencies."

Touched, Jack's voice broke as he tried to reply. He knew they couldn't afford it. He quickly coughed to try and cover it. "Can you look after Skittles for me?" he asked, scared Mum would use his absence as an excuse to get rid of his cat.

"Of course," Ryan said. He patted his belly, one of Skittles's favourite sleep spots. "You know that cat can rest safely around me."

Mum returned with a twin gripped in each hand. She looked at Jack, her eyes red-rimmed. She hugged Jack so uncomfortably close, he squirmed for air. "Bye, Puppy." She hadn't called him that since the twins were born. "I'll miss you."

Jack stiffened. Did that mean she was going to miss having an unpaid babysitter and cleaner?

"Jack, if you want to come home, you just let us know," said Mum. "Okay? Just call me."

Jack ripped himself from her tight hug. If she wanted him to be independent, then he would be. There would be one less mouth for her to feed, and the twins could soak up every single bit of attention.

"I love you," Mum said.

"I think that's the boarding call," Jack muttered, grabbing his carry-on.

It had been a long journey since then, and he had been drifting in and out of sleep.

Jack woke again as the plane bumped and rattled over a dirt airstrip, limping to a stop next to a rusting iron shed.

A few people sheltered in the shed's meagre rectangle of shade. They ambled out as the propeller blades slowed to a stop. Cardboard boxes wrapped in tape were hoisted out of the plane and put onto the trays of a couple of waiting trucks.

Jack took a deep breath before swinging his bag onto his shoulder.

A tall man with blue eyes squinting from beneath a battered hat approached, with the slow, steady gait of someone who was never in a hurry. "Jack?"

Jack nodded, suddenly horribly shy.

"It's good to see you. You've grown so much."

Jack didn't know what to say. If his dad could have made even the occasional effort to see him over all these years, maybe he would have known how much Jack had grown. "You too," Jack mumbled. He started to explain. "Sorry, I mean, it's good to ..."

But this stranger, his father, was laughing. "You don't have to apologise, Jack. I understand this might feel odd to you. It certainly does to me. But I'm glad you asked to spend some time with me."

Is that what Mum had told his dad? Not that Jack was coming because he was difficult and had too much attitude now that he was nearly a teenager?

Jack's dad whistled. A dirt-coloured dog seemed to emerge from the surrounding desert, bounding over. "Meet Dusty," his dad said, meeting Jack's eyes. "I know what you're thinking: could I have named her something a little less obvious?"

Despite himself, Jack grinned. It was exactly what he had been thinking.

"Dusty is all lick and wag; she wants to be everyone's best friend."

As if on cue, Dusty leapt, wagging her tail wildly, and planted her front paws onto Jack's chest as if inviting him to dance with her.

His dad hefted Jack's suitcase from where it had been deposited next to the airport shed. "The only thing Dusty absolutely hates is reptiles," he added, "which is a bit ironic given what I study for a living."

Throwing the suitcase next to the taped cardboard boxes on his truck tray, his dad whistled for Dusty to jump up. "I guess it's now or never to show you what home looks like, but you might find that my place is different from what you're used to."

They were silent in the truck as they bumped along a perfectly straight stretch of ribbed dirt road, leaving an orange plume of dust streaming behind. It was as if his dad did not feel the need to speak, or maybe it was just too loud with the engine and the jostling and thumping. He didn't turn the radio on either, just kept his eyes fixed on the horizon.

Jack wondered if it would be rude if he listened to music on his headphones. He was still dazed by all of this ... nothingness. His dad literally seemed to live in the middle of nowhere. If Jack did just *want to come home*, as Mum had said, it wouldn't be as easy as catching a train or bus.

Jack's eyes were hungry for trees, for buildings, for the familiar. Instead, there was only this flat stony ground, with the occasional brave stalks of bleached-looking weeds and thorny tangles clawing out from cracks in the baked earth. He saw a lizard scuttling. Dusty spotted it, too, barking madly and flinging herself off the truck's tray in chase.

"Um ..." Jack couldn't bring himself to call this man *Dad*. It felt weirdly familiar – too close. "Dusty jumped off. Shouldn't we stop for her?"

"She knows where home is. It's not far." As if he'd guessed what Jack was thinking moments before, Jack's dad added, "I've been contemplating what you should call me. The fact is, we don't really know each other. So, I will call you Jack, and how about you call me what everyone else around here does? Rocky."

"Like 'Rocky' the boxer from those old eighties movies?" Jack asked.

"No, like a guy who likes to spend his life looking at rocks."

Jack nodded. Dusty. Rocky. He wouldn't be surprised if they got to the house, and it had one of those little signs with the property name carved on it: *Housey.*

The truck slowed to a halt, and then Jack glimpsed his new home. Except, it wasn't a house.

Four

The Grand Tour

Even the word "shed" would be generous.

"It probably doesn't look like much to you," Rocky said, "but it's a whole lot better than where I stayed in Mongolia. I didn't even have a tent there, just a mattress pad made from felted camel hair."

An iron roof overhung some posts, but there were no walls on three of the sides. A patchwork of scuffed and ripped tarps formed the floor, while a sink stood against the single wall, its draining rack sporting a shallow tray covered by a green carpet of sprouts. Next to this, an old bookshelf housed glass jars of rice and beans. The rest of the kitchen consisted of a bench with a single gas burner ring, a gas bottle underneath and plastic tubs stacked on the other side.

In the centre, a long rectangular table was crowded with wooden trays and small mounds of rocks. Some seemed to have been hastily swept aside to create a space just large enough for two people to set down plates and sit opposite each other in mismatched chairs.

Dusty arrived, panting, and tore through an assemblage of junk to leap onto a collapsible camp bed in one corner. In the opposite corner, Rocky had rigged up a kind of cubby made from canvas, half screening another camp bed with a sheet and an unrolled sleeping bag on it.

Rocky gestured to the cubby. "I know you would have had your own room back with your mother, so I wanted to give you some privacy."

Jack stared. His own room? When Mum and Ryan had moved into a downstairs bedroom to be closer to the twins, he hadn't just had a room, Jack had practically had the entire upper floor to himself.

Rocky gestured to a large storage tub with a clip-on lid. "You might want to put anything you value in there. There's a type of rat out here that will eat through anything." He flashed Dusty a disappointed look. "Madam here would be happy to welcome rodents in, show them around and reveal all the places I stow my treasure."

Rocky caught Jack's look of disbelief. "Not all treasure is a pirate's chest of rubies and gold. I have found some fossils in my time ..." He reached into the neck of his long-sleeved shirt to reveal a thin leather strap. From it dangled what looked like a small rock. "Do you know what this is?"

Jack shook his head. He still hadn't really recovered from shock. This place was a dump.

"It's extremely rare," Rusty said, handing the pendant to Jack. "And another part of a 66 million-year puzzle. Which came first: the dinosaur or the chicken egg? Do you know that some dinosaurs had feathers? Some palaeontologists believe that since the meteor struck, today's birds are some of the only dinosaurs left."

Jack gazed at the faint imprint of what looked like the fanning lines of a feather etched into the rock. "What would this be worth?" he asked Rocky.

Rocky smiled and reached to take it back. "It's priceless," he said. "Now, you must be tired. We'd better finish the grand tour." Rocky wove through stacked boxes, tools and piles of books and papers. "The shower's outside."

To Jack it all looked like one big "outside". Rocky pointed to a post with a silver bag hung from a hook. "The sun heats the water to almost boiling during the day, but it gets cold out in the desert at night, so it's nice to have a warm wash before it gets too chilly. We use the sun out here to power the lights, too. These lamps," he said, indicating a motley row of lanterns, "are all solar charged."

Rocky pointed to two pieces of sheet metal forming a tent shape some distance from the shower. "That's a drop-pit toilet. If you need to go at night, just make sure you take a flashlight. Or call Dusty to go with you. Snakes don't like the cold, so sometimes they seek out warmth where they can find it."

Jack's insides constricted. He thought longingly of the clean white tiles and porcelain of his bathroom at home. How long could he go before having to use a snake-infested drop-pit toilet?

"As you can see, it's basic," Rocky continued, "but when I'm out here on a dig, I like it that way. Most of my days are spent out digging or sorting anyway. And at night, the stars are on show. Anything I haven't mentioned?"

Jack tried to keep his voice steady. "I guess you don't get Wi-Fi out here?"

Rocky shook his head, looking almost proud. "Not here, but you will be able to use the internet at your school. I have a satellite phone, but it's expensive to use, so I mostly reserve it for emergencies. I understand you must be missing your mum and brother and sister, though, so if you'd like to use it, you can call them to say you're okay."

Jack shook his head, not trusting himself to speak in case he started to cry. No way was he going to let Mum know that he'd landed in a nightmare and was already desperately homesick.

Rocky tore the tape off one of the boxes the plane had delivered and rummaged through it. He pulled out a couple of brightly coloured packets. "It's been a big day and you'll be going to school tomorrow – your mother insisted that you don't miss any schoolwork. So how about we have a lazy dinner? A cup of soup and bread as fresh as we can get, straight from the plane. We can put some sunflower sprouts on top. Then donuts for dessert?"

Jack nodded wearily, suddenly wanting, more than anything else, to lie down on that dinky little camp bed in the canvas cubby and sink into sleep.

Five

The One-Room School

Jack spluttered awake to Dusty enthusiastically licking his face. It seemed Dusty was going to be his alarm clock now. Which was worse: sore ears from the twins' constant noise or being drowned in dog drool?

Jack lay cocooned in his sleeping bag as a wash of honeyed light slipped into the shed, turning everything a rich gold colour. Outside, the deep turquoise sky was still pinpricked by a few stars, until a huge orange sun crept over the horizon. At home, there were so many buildings that by the time the sun climbed past them all, it was already a pale gold circle in a bright sky.

Rocky lounged in a chair facing the east, nursing

a cup of steaming tea. "I never tire of this view." He turned to Jack, who was struggling to rise from where Dusty was pinning him in his sleeping bag. "Di and Carlo from the cattle station have offered to give you a lift to school," Rocky said. "We're not entirely alone here; our neighbours are just a little way down the road. They have a bunch of kids, so many I can't keep count, but one of them is bound to be around your age."

"What am I meant to wear to school?" Jack asked, suddenly nervous. "Is there a uniform?"

"No uniform, just your usual clothes." Rocky fumbled with a paper lunch bag, handing it to Jack with a shy smile.

Up until a couple of years ago, Mum would have made egg and lettuce sandwiches, with cubed melon and lychees packed neatly into the lunch box compartments. Or sometimes, Jack was allowed to buy sushi or pizza at the canteen.

Jack peered into the bag. Inside was an orange, a small packet of chips and what smelled like a peanut butter sandwich.

"I hope I got it right?" asked Rocky. "I've never done this before."

Jack nodded, strangely touched. Rocky might have

been completely off the scene for most of Jack's life, but he also hadn't asked for his stranger-of-a-son to suddenly arrive from out of the blue to live with him.

An old minibus rattled up the dirt track. Jack took a deep breath.

The woman who was driving beamed, instantly deepening the fans of smile lines crinkling her eyes. She had short brown hair with a single purple streak, and a flower tattoo on her shoulder.

"Di, this is Jack. Jack, Di," Rocky said.

"Great to meet you, Jack. Our kids have been pestering us ever since Rocky said you were coming." Di gestured towards the back of the minibus, which was chaotic with kids of seemingly all ages unbuckling their seatbelts to bounce between the seats now that the bus had stopped. The oldest-looking one appeared to be about Jack's age. She stayed put, cradling a black chicken in her lap.

"That's not true," the girl with the chicken sang out. "You're the one who couldn't wait to see what Rocky's son would be like."

Di shrugged, as if in defeat. "What Blossom says is true. I've only ever known your father to be a solitary man."

"If you search for him online," Blossom interrupted, "he's written three books and he's practically famous. He's a doctor. I'm going to be a doctor when I grow up and work for Médecins Sans Frontières."

"Rocky isn't a medical doctor, he's an academic," Di corrected.

Blossom ignored her. "Do you know what Médecins Sans Frontières is?" she asked Jack. Without waiting for his reply, she answered for him. "It's French for 'Doctors Without Borders'. They go to countries that have wars going on and help heal people for free."

Jack was impressed. He'd barely thought about what he was going to do when he grew up. Sometimes he thought he'd like to design computer games or be a professional gamer, but every kid thought that.

"Okay kids, buckle up!" Di shouted over the din.

Blossom shifted, with a squawk of protest from the chicken, to make room for Jack on the bench seat. "What's it like where you come from?" she immediately asked him. "Do you miss it? Why did you come out here? Are you going to look for fossils like your dad does?"

Di called back over the younger kids. "Blossom, enough with the hundred questions. Give the poor kid a break, he only just arrived."

Blossom rolled her eyes. "This is Henny Penny," she said, introducing Jack to the chicken, who eyed him suspiciously. "It's a clichéd name, but what would you expect from a little kid? What would you have called her?" she asked Jack.

"I don't know. Um – Sasha?" Jack's friend, the real Sasha back home, would have tossed her glossy golden hair, offended.

Blossom gave a brief nod. "I would have called her Ava because 'avian' is the scientific name for birds."

Jack thought about what Rusty had said last night about some dinosaurs having feathers. It had taken 66 million years for a giant T-Rex to turn into a lap chicken.

The bus pulled up next to a freshly painted timber building. It had an ornamental facade and old-fashioned curly letters announcing that the school had been established almost a century ago. Behind the building was a portable classroom. Blossom's brothers and sisters tumbled out of the bus and raced towards a brightly coloured swing, slide and climbing frame.

Blossom, regally bearing the avian Henny Penny, ignored the swings, now swarming with her siblings, as if too old for such childish things. She beckoned for Jack to follow. "You can sit next to me today."

School turned out to be a large single classroom in which the teacher, a red-faced man called Mr Azzopardi who periodically pulled a handkerchief out of his pocket to wipe his sweating forehead, assigned different lessons across the year levels. Mr Azzopardi marked the roll, making no comment about Henny Penny's presence, only pausing to ask the others to welcome their new student.

It was a tiny one-room school in the middle of nowhere: the opposite of Jack's three-storey state-of-the-art school back home. But Jack was surprised to discover he wasn't ahead in his classwork like he had assumed he would be. The work that Mr Azzopardi assigned to him, Blossom and the three other kids in their year level was challenging.

The best thing was having access to the internet. There was no point in most of the kids out here having phones, as mobile phone coverage was either patchy or non-existent. But once they had finished all their assigned work, students were allowed to have free time on their laptops. While Blossom scrolled through online videos that demonstrated the differences between applying a tourniquet and a compression bandage, Jack logged into his email.

There was one from Mum in his inbox with the subject line *Missing you already xoxo.* He ignored it. Instead, he emailed some of his friends to tell them that he couldn't send or receive texts or message them as usual, only email.

He was stuck for what to write to them. Maybe: *My dad's place is a total tip with no internet and snakes in the toilet and there are no trees or anything green around here at all.* Jack tried to angle the computer so Blossom couldn't see, and instead he wrote: *My dad has a cool dog, and he finds fossils that are worth more than diamonds and gold.*

There was no bell, only a clock on the wall to let them know when it was morning tea or lunch. At morning tea, Jack found, to his quiet delight, a toilet block with real toilets that flushed, clean and still fragrant with toilet cleaner.

In the afternoon, the sudden hectic scuffling and stuffing of things back onto shelves and into bags alerted Jack to home time.

As Di's minibus rolled into the yard, Mr Azzopardi stopped Jack to congratulate him on a good effort for his first day. Blossom tugged at Jack's arm, asking him, "Do you want to come round to our place and ride horses this afternoon?"

Six

A Totally Different World

So long as Jack could remain good-humoured about Blossom's little brothers and sisters treating him like a slapstick rodeo clown brought in especially for their amusement, attempting to ride a horse was a lot of fun – even if it was their oldest, slowest horse, as Blossom pointed out. Learning how to milk a cranky goat was pretty interesting, too.

When Rocky came to pick up Jack, just as it began to get dark, Di and Carlo, a shaggy, bear-like giant, convinced them to stay for dinner. The kitchen was already filling with the rich aroma of onions and grilled burger patties.

Unlike Rocky's place, Blossom's family lived on a cattle station in a house with the right number of walls. There was solar power for the whole house, so all the lights blazed brightly. "We might be short on nearly everything else," said Di, "but one thing we've got plenty of here is sunshine."

"And fossils," Rocky added. He pointed to Henny Penny, plumped down on Blossom's lap, and shook his head. "It's still bizarre to me to think a raptor could become *that*."

"Leave my chicken alone," said Blossom. "Come on, Jack. I need you to lend me a hand ... literally." Then she pulled Jack away from the table and into the living room to practise her bandaging skills. Blossom pulled out a clean bandage and wrapped it as tightly as she could around the top of Jack's arm, muttering instructions to herself the whole time.

"Hold still," she demanded.

"Ouch!" Jack protested.

Blossom shrugged. "A tourniquet has to be really tight to slow the blood flow. Otherwise you could die."

"Only I'm not bleeding," Jack pointed out.

He sat patiently while she unwound the bandage and applied it again. "This one is a compression bandage," Blossom explained. "It should be nice and firm but not too tight. It's what you use for a snake bite. You have to wrap the bandage a bit above and below the bite."

Jack thought about how Rocky's outdoor toilet was a possible snake pit, and shuddered.

"Did you know the venom doesn't go into your blood until it's gone through your lymphatic system?" Blossom asked. "That's why you put on a compression bandage, to stop the venom from moving through your system. You shouldn't move, either, because muscle movement makes the lymphatic system pump fluid around until it gets into the bloodstream."

Jack nodded, trying to take it all in. Sometimes listening to Blossom was like being unable to press pause on an audiobook.

"If you could travel to another time and then come back again, what century would you choose?" Blossom asked.

Jack considered. Rocky would choose to go back to the age of the dinosaurs. But for himself? "I think I'd like to go into the future, maybe five hundred years from now, and see what kind of technology we have. Maybe we'll have robots and wear contact lenses that show us different scenes while we're walking down the same street."

"I'd like to go into the future too," Blossom agreed. "But mostly so I could see what medical discoveries have been made. Maybe by then we will have cured cancer."

Jack smiled. Blossom could be bossy, but she had a kind heart. One of her sisters had gravely explained to him that Blossom took Henny Penny to school each day because Henny Penny had been so badly attacked by the other chickens that half her feathers had been pulled out – and Blossom wasn't having that!

Jack and Rocky didn't get back home until late. For a while, Jack lay on his camp bed, listening to Rocky quietly plucking at a battered guitar. It had been a strange day, unlike anything Jack had ever experienced before. It was as if he'd been dropped into a dream – a totally different world with different characters and different ways of doing things. Jack couldn't honestly say that he'd enjoyed it all, but he hadn't hated it either – making friends with a girl who brought a chicken to school, learning how to ride a horse bareback, even being kicked by an irritable goat. Then there was that almost crazy joy he had felt upon realising there were toilets at school, in cubicles, and they flushed!

Jack's body was weary from all the romping around he'd done with Di and Carlo's kids. Dusty mooched over to lick his cheek on the one spot she could access through the cocoon of Jack's sleeping bag, and Jack pulled the dog closer, burying his fingers in her thick wiry fur as he drifted into sleep.

Seven

Stars and Time

School over the next few days was more of the same. Blossom had clearly claimed Jack as her friend, but graciously allowed him to play softball with the other kids their age while she tended to Henny Penny. There were two boys, Luke and Sanjay, and another girl, Talia, who were equally friendly and clearly excited to have another kid their age to help cement their authority as school seniors.

During his free time, Jack responded to emails from his friends Sasha and Bryson. But it wasn't like messaging them – it felt a bit forced. Nothing much had changed in their lives during the course of a week, while Jack's life had been turned upside down.

There was an email from Ryan: *Hey Buddy, how's it going out there? The twins keep asking for "Dak", and Skittles is sticking to me like chewing gum ...*

There were three more emails from Mum, with the most recent titled: *Jack???* He didn't bother opening them. She wanted him to miss home, to miss her. Two could play that game.

After school each day, Di, Carlo and "the full catastrophe", as Rocky referred to the kids, seemed happy to have Jack join them. Rocky had explained that it suited him – he could stay out on the site longer in the cooler part of late afternoon, but still be back for Jack before dark.

Jack was relieved to be surrounded by noise and company at the cattle station. The quiet at Rocky's place was unnerving. There was nothing for him to do there. There were books on fossils and a single scuffed pack of playing cards. Apart from that, there were none of the usual forms of entertainment – no TV or computer or phone, or even board games. Rocky sometimes played older style riffs or classical music on his guitar, but most of the time he seemed to enjoy the lack of noise.

At night, Rocky lit a fire with sticks and tussocky grass. He pointed through the blue smoke swirling towards the stars and told Jack about some of the constellations. "The First Nations people of this land have been doing this for thousands of years. It's said that when they sat around fires at night, the people looked up and listened to stories from the stars. These days, though," Rocky added, "a lot of us newcomers can't seem to listen properly, so we have to tell the stars our own stories."

He mused a while, then added, "In any case, we're all part star. The iron in our bodies, yours and mine, originally came from baby stars."

A shooting star arced across the horizon, burning a long, bright trail that gradually faded into the velvety dark. Mum would have said: "Make a wish, Jack."

"Meteor," Rocky drawled, matter-of-fact. "It's just a broken-off fragment of an asteroid that's entered Earth's atmosphere and burned up."

Rocky continued, "You know, it's thought that the dinosaurs were wiped out by part of an asteroid that hit Earth 66 million years ago? That, and possibly supervolcanoes erupting. Some scientists think the soot and debris blocked out the sun. It created a night that lasted for two years, turning things way too dark and cold for most creatures to survive. Imagine that – two years of pitch black."

Jack tried to imagine. For him, night had mostly just been a time to have dinner, go online, watch something on his phone and catch some sleep before starting the next day.

"You might wonder why I'm out here, Jack," said Rocky. "When I was your age and younger, I mostly just didn't understand other kids. It would wear me out. All I wanted to do was play with my toy dinosaur collection. Things would feel too busy and overwhelming, and I often preferred to be on my own."

Rocky poked a stick into the fire, sending a shower of vivid orange sparks into the sky. "Time moves slower here. You'll get a better idea when I take you to the dig this weekend. When you're puzzling together the stories and remains of creatures who lived millions upon millions of years ago, you're working with a different timescale. Geologists call it deep time. It's all about understanding that the scale of planetary events is so much vaster than the petty plans of us humans."

Rocky pulled the fossil pendant from around his neck and handed it to Jack. It was still warm from where it had lain against Rocky's chest. Jack ran his fingers over the rough serrations and edges, the fine indentations of feather turned to rock.

"Dinosaurs ruled this planet for over 165 million years," Rocky continued. "Humans for only a fraction of that time, and even then, what we think of as modern humans have only been around for about a few hundred thousand years. Out here, I experience actual days and nights – not when someone switches on a light to create a fake day at midnight. It's how most living things, humans included, would have lived up until a century or two ago. Deep time can make many of our human concerns seem so trivial and fleeting."

Jack wondered if Rocky had thought being part of his own son's life was a "trivial" human concern. Why were fossilised animal bones so much more interesting?

So far, it had been mostly Rocky who did the talking. He didn't ask many questions. Not like Mum, who, before the twins, had always been chatty and curious. Or even Blossom – as much as she liked to tell Jack about all the new things she learnt, Blossom also fired off a thousand questions about what his life was like back in the city, and then actually listened to his replies.

Rocky met Jack's eyes across the fire. "Bringing you into the world made me think more closely about deep time – the past, present and future. As someone more poetic once said: you're my arrow into the future. Maybe one day you'll have children, and they'll get to sit around a fire and look up at the stars and realise that we're all millions – billions – of years old."

Eight

Piecing Things Together

As the truck rattled along two thin track marks in the dirt, a long, low ridge rose from the otherwise flat plains, like a beached whale turned to stone. Apart from the ridge, it all looked identical to Jack. "How do you know where to look for fossils?" he asked.

Rocky shrugged. "Sometimes it's a hunch, but this time it was actually Carlo who said he thought he'd found something interesting. This ridge is on Carlo and Di's station. I've spent hours out here looking for bits of exposed bone. If the team's out here and we find something, we cut whole slabs of rock away to try and minimise damage to the specimens."

As they climbed out of the truck, Rocky gestured to the expanse. "Strictly speaking, I should have other people with me. But there's a section I've been working on for months by myself. Besides," he said, awkwardly putting his hand on Jack's shoulder, "I've got a new helper, now. We can consider ourselves lucky that we're not back in the terrain of poor old Mary Anning."

"Who's that?" Jack asked, scuffing his sneakers into the dirt.

"Mary Anning lived a couple of hundred years ago in the Lyme Regis area in England, on the Jurassic Coast, before people had a good understanding of what fossils really were. She found an entire ichthyosaur when she was twelve years old. It was dangerous work though, she was always at risk of being drowned by the rising tides, or sudden cliff collapses. Her dog was killed right next to her by a rock fall."

Catching Jack's expression, Rocky added, "Don't worry, the main rule is to remember to keep your hat on! The only thing you're likely to die of on this dig is heatstroke."

Or boredom, Jack thought.

Rocky gave him a tool that looked suspiciously like a toothbrush and showed Jack how to painstakingly brush dirt from the small ridges in a possible fossil. To Jack's surprise, after initially feeling frustrated and overwhelmed at just how much desert there was to brush away with a *toothbrush*, his mind quietened and he drifted into memories and thoughts that became increasingly spacious and calm. Was this what Rocky had meant by the feeling of deep time?

They worked peacefully alongside each other. The hypnotic effect of softly brushing with small, neat strokes meant Jack was surprised when Rocky announced it had become too hot, and it was time to go home for lunch. Where had the hours gone?

Jack had the same feeling later, when Rocky showed him some of the rocks in the trays in his collection. "Usually, I'd have someone we call a preparator doing this work – scraping and brushing out the grit and piecing together bits of broken bone. But out here, I do it myself. It's like putting together a puzzle millions and millions of years old."

He demonstrated with one of the rock shards and handed it to Jack, who was again surprised to find the task oddly soothing.

At one stage, the thought of his friends, laughing and teasing each other through their headsets while they teamed up in an online game, flashed into

Jack's mind. He couldn't begin to imagine explaining how it was possible to enjoy piecing together fossils almost as much as getting the highest score.

That night, Jack was startled awake by a ringtone blaring through the darkness. He heard Rocky lumber though an obstacle course to answer the satellite phone. "Hello? Is there anything wrong?"

There was a pause. Rocky sounded surprised as he said, "No, that's okay, but I think he's asleep. Do you want me to wake him?"

There were more scuffles and a yelp from Dusty as Rocky tripped on his way to Jack's corner. *"Jack,"* Rocky whispered loudly. *"Jack! It's your mother!"*

Jack would have found it comical if he wasn't so confused. He didn't know if he wanted to talk to Mum. She would have known how Rocky lived, the fact that he was in the literal middle of nowhere, and she'd still sent Jack here anyway.

Jack lay still, pretending to sleep, but Rocky wouldn't give up. "Your mum wants to speak with you."

Sighing, Jack reached for the phone. He fake yawned. "Hello?"

"Jack? How are you? You haven't replied to my emails," said Mum.

"I've been busy."

Mum was silent. She would know what this place was like. No way could she imagine anyone would be busy out here. He heard her swallow. "We've been missing you. Daisy keeps looking under beds and in wardrobes to see where you're hiding."

Something in Jack's throat caught. When they weren't screaming, the twins could be very cute. He imagined little Daisy's dark eyes widening as she bent to look under the beds or slid open wardrobe doors. It made him long to be able to spring out, snatch her up and throw her into the air shouting "Surprise!" while she dissolved into a fit of giggles. Or to have Danny insist on riding Jack's feet like a giant, crowing "Dee di do dum."

But what about Mum? Had she been missing him?

"How is everyone?" he finally asked.

Mum gave a tired laugh. "The twins have been non-stop."

Jack closed up again. It was always all about the twins. Before the twins, she had always asked him how his day at school had been, and then actually listened when he answered.

Admittedly, Rocky didn't ask many questions

either, but Jack didn't expect Rocky to ask questions – that was the difference.

"It's been pretty busy out here, too," Jack said. "It's great. I'm learning all about dinosaurs and fossils." Then, hearing Danny starting to whine in the background, he added, "It's really peaceful and quiet."

"Jack ..." Mum began, "I don't want you to think that ..."

Danny's whine became a furious howl for attention.

"Ryan!" Mum called. "Ryan, could you please take him? Okay, just forget about it, then. Listen, Jack, I'm really sorry but I have to go. I'll try again soon. I love you."

"You too," Jack replied begrudgingly, but the line was already dead.

Nine

Being Where You Are

Over the next few weeks, Jack turned Mum's phone calls over and over in his mind, as if holding up a fossil, examining its different angles and facets, trying to get a clearer picture. Every time Mum rang, he knew it must be costing her a small fortune. Jack would hear the twins or Skittles yowling in the background and feel waves of homesickness. He missed the familiar smells of their kitchen – a real kitchen, with no desert rodents. And his own room and bathroom with a toilet that had a closing door and no threat of snakes. He hadn't actually seen a snake in the drop-pit toilet yet, but it always felt like they were lurking.

As much as Jack had sometimes wanted to take the twins back to the hospital for a refund, they'd then do something so heart-meltingly cute, he would have been embarrassed for his friends to see how gooey he became. Jack missed his friends too, although Blossom and her brothers and sisters were fun to hang out with, and if he had to weigh it up – riding horses was a lot more enjoyable than pressing buttons on a controller.

It was calm here, and genuinely interesting spending time with Rocky, who treated Jack like an equal. He wasn't really like a dad, not like Carlo or Ryan. He didn't ever hug Jack like Ryan did – ruffling his hair and slinging an arm around his shoulder. But given the situation, Jack would have found it more awkward if Rocky had been affectionate.

Rocky was respectful of Jack's privacy, and had proposed a simple system of them taking turns clearing up after meals. The peanut butter sandwich and orange from the first day had been a one-off – after that, Rocky had told him to "feel free to forage and take whatever you like."

Rocky didn't make Jack do his homework, but since there wasn't much else to do, Jack did it anyway, usually after being dropped off by Blossom on her horse, Shadow. At night, Rocky and Jack sat around the fire, or at the table with their solar lanterns, sorting bits of fossil from Rocky's daily forays to the ridge. Rocky never talked down to Jack or ordered him around. He mostly just explained things.

Mum had once told Jack that when she'd moved to a new country, she'd been too excited to be homesick. She would feel only an occasional wave of missing people and places, sights and smells. Mostly, though, she just got busy with her new life. "I have a talent for being where I am," she'd told him as they shuffled forward in the queue at Jack's favourite Chinese bakery after his weekend football match.

The sun sank and a soothing rush of wind cooled Jack's face as he sat contentedly behind Blossom on Shadow, trotting towards Rocky's shed. Jack wondered if perhaps he also had a talent for being where he was.

Ten

To the Ridge

For the first time ever, Dusty didn't bound out to greet them with enthusiastic barking. Neither did Rocky amble out after her, wiping his hands on dirt-creased trousers from where he'd been sorting fossils.

The place was eerily quiet. Lengthening shadows seemed to swallow the shed's interior, and Jack realised something else was missing. The truck was gone.

As had become typical, Blossom pointed out what Jack was thinking. "That's weird, Rocky is *always* here when we arrive."

"Maybe he had to go somewhere?" Jack suggested. "Like out to the airport or to a shop or something."

Blossom raised her eyebrows. "Planes only come to the airstrip on Sundays and Wednesdays." As for shopping, they both knew the nearest shop was a three-hour drive away. "He would have left you a note, or let my mum and dad know."

Jack climbed down from Shadow, removed his helmet and dashed into the shed. There was no obvious note on the table or near Jack's bed. He came back out shaking his head.

"Do you think he might have had an accident?" asked Blossom.

Jack sighed. First aid was always on Blossom's brain. "Maybe his truck broke down," he said.

Rocky would have gone out to the ridge again today, like always. But he was always back in time for Jack. If something had happened to the truck, it would be too far to walk back in the blistering heat. "Maybe he got stuck," said Jack.

"We should go and try to find him," Blossom replied.

"It's too far away."

Blossom grinned. "Not if we ride Shadow. We can go cross-country while there's still some light."

Jack refastened his helmet and clambered back up, and Blossom gave Shadow a command so that they were soon galloping their way towards the distant, blue-shadowed ridge.

As the ridge loomed closer, Jack strained to see through the darkening twilight. Finally, he made out the blocky shape of the truck. They soon reached it, and wild barking filled the air. Dusty came dancing out towards them, yapping frantically.

Jack and Blossom jumped off Shadow as Dusty leapt and tugged at Jack's sleeve, as if to pull him along, further and higher up the ridge. It was steep enough in some sections that Jack and Blossom had to use their hands and feet to climb.

Around halfway up the ridge, they found Rocky lying on a ledge. "Rocky!" Jack shouted. "Dad!"

Rocky's eyes fluttered open. "Good to see you both here," he said.

Jack's heart thumped as he scrambled to kneel next to his father. "What happened?"

"I was just about to go home when I had a hunch there might be something of interest up here." Rocky managed a weak smile. "Turns out when I told you there wasn't anything too dangerous to worry about, I forgot about snakes. I put my hand in the wrong place."

Jack now saw that Rocky had his shirt tightly wrapped around his hand.

"You have to stay still!" Blossom commanded.

"That's why I'm lying here," Rocky agreed. "I figured if I tried to walk back to the truck the venom might kill me sooner. I hoped that when you saw I wasn't home, Di and Carlo would raise the alarm. There's a first aid kit in the truck cab."

"I've got a bandage with me here," Blossom said, unable to disguise the triumph in her voice.

While Blossom carefully wound the bandage around and around Rocky's hand, from his fingers to halfway up his arm, Jack felt his stomach clenching with fear. This wasn't an internet video. It was real. His dad had been bitten by a snake and he had lain here for hours, alone, not knowing if anyone was going to find him in time.

The moment she finished, Blossom said, "I'll have to get back to my mum and dad. They're going to have to radio for a plane."

She looked at Jack, but before she could bossily state the obvious, Jack said, "I'm going to wait here, with Dad."

Eleven

Acts of Love

What could have been another peaceful evening under the stars had become a nightmare.

"Jack?" asked Rocky.

"You don't have to talk," Jack shushed his dad, worried that even the effort of talking might make the venom take effect faster. He shivered as the heat of the rock beneath them began to evaporate in the desert night.

"I've done the right thing, Jack – barely moved a muscle. I don't know how bad it is. I felt the fangs strike but I'm not sure how much venom got in. Anyway, it's given me time to think about things I'd like to say to you."

"Maybe you should tell me when you've been

treated and you're safe," Jack said. If he acted like everything was going to be okay, then maybe everything would be.

Rocky smiled. "I think I'd like to tell you now – about how you became my arrow into the future. I want you to know that I tried. Your mother had come from China on a student visa. She had grown up in one of the world's biggest cities and dreamt about what it would be like to be somewhere with no people, no buildings, often even no roads. She dreamt of living in a tropical jungle where snow is something people only read about."

Rocky sighed. "I like solitude, but I still enjoy some human company when it turns up. I enjoy *your* company. And when your mother came for her big adventure on a dig, she had a way of listening – I could tell she wasn't just pretending to be interested, she really wanted to learn. We fell in love."

Jack tried to imagine his parents as they might have been back then. He pictured the way Mum used to tilt her head when she was listening to him tell her about something that happened at school, as if it was the most important thing in the world. He could see why Rocky might have been drawn in by that kind of attention.

"When we found out you were coming along, we tried to make it work," Rocky continued. "At first, she stayed out at the field dig with me. But what had seemed romantic at first – the candlelight and endless stars – became monotonous. Your mother is a city girl. She's used to having lots of activity and people surrounding her. What was good for her was torture for me, and vice versa."

A night bird called out. Dusty put her head up and then nuzzled back against Rocky's side.

"I followed her back and tried to do the city thing for her, for you, but I was miserable working in an office and couldn't hide it. We started to argue. Giving me the freedom to leave was your mother's act of love."

Jack contemplated the way Mum had never said anything mean about Rocky. It was as if she'd put him in a box that wasn't to be opened.

"She gave me freedom, but her terms were tough," Rocky told him. "She didn't want you being batted back and forth between us. She wanted to give you something solid. I tried coming to see you a few times. I took you to the museum."

Jack had a glimmer of memory – boredom and discomfort, looking at an endless array of rocks in glass cabinets.

"She said you found the visits unsettling," Rocky continued. "Then she met Ryan and it seemed as if he was the perfect dad for you. I know you're angry with her for not understanding you right now, and you've been brave and strong coming out here to stay with me. But I think you miss her, that you miss your family."

Dusty stiffened and her ears pricked as a faint whirring grew louder. A beam of white light swept down from the sky towards them. A helicopter.

Jack leapt up, frantically waving his hands. "Over here!"

Through a haze of panicked relief, he only just heard Rocky add, "If I make it through this okay, you'll always be welcome to come back and stay, but telling you this is my act of love for her, and for you – it's time for you to go home, Jack."

Twelve

Arrows into the Future

Jack blinked back tears as Mum folded him into her arms and hugged him tightly, as if she'd never let him go. "I am so sorry, Jack. So very sorry. It was wrong of me to send you away like that. I just felt as if I was going crazy."

Ryan, wearing a superhero t-shirt, ambled back to them in the carpark. He was beaming and juggling five dripping ice-cream cones.

"Rocky said that I can go back and visit when he gets better," Jack said.

Mum and Ryan exchanged glances.

"It's going to be a long road for him," Mum said. "He's still got quite a bit of recovering to do. And in the meantime, it looks like we have another addition to the family you'll be spending time with." She glanced at the back seat, to where that addition sat wedged between the twins' car seats. Dusty gave an exasperated whine.

Jack pulled himself from his mother's arms and rapped on the window. "Don't pull her tail!"

To his astonishment, instead of telling him off, Mum rolled her eyes. "Sometimes I wonder what I did to deserve those two."

"They're your arrows into the future," Jack said.

In answer to her raised eyebrow, Jack found his fingers wandering to the fossil pendant beneath his t-shirt. Rocky had told him to hang onto it until they saw each other again.

Jack explained, "It was something Rocky said, about deep time – how we continue to live on even though we only have such short lives. The twins are your arrows into the future."

Mum smiled. Her eyes softened as she regarded him with that loving gaze he remembered from way back, before the diabolical duo had come along. "And you are, too, Jack."